PSOAS MAJOR MUSCLE THICKNESS

Dr. Veronika Vasiliev

Kelsiehanson Verlag

DEDICATION

To My Wonderful Readers

CONTENTS

ACKNOWLEDGMENTS

Thank you to everyone who strives to grow and help others grow

1 ABSTRACT

The Psoas major muscle (PMM) is a long, thick, fusiform muscle and one of the muscular components of the posterior abdominal wall. The aim of this study was to assess the psoas major muscle thickness in apparently healthy adults in Port Harcourt metropolis. Two hundred and ten (210) apparently healthy volunteers aged 18 years and above were recruited. A high frequency (7.5 MHz) linear transducer of a 4D LOGIQ P6 General Electrics Ultrasound machine was used to measure the muscle at the point of maximum thickness. Data was analysed using SPSS version 21.0. Pearson correlation analysis was used to test for the correlation between PMM thickness and Age, Body Mass Index, Height and Weight.Statistical test was considered significant at $p \leq 0.05$ at 95% confidence interval. The mean PMMT was34.28±5.15mm and correlated with age ($r= 0.304$, $P<0.05$), height ($r=0.234, p=0.001$), weight ($r=0.535$, $p=0.000$) and BMI ($r=0.443$, $p=0.000$). The PMMT was greater in males with a mean difference of 3.08mm. A mean difference of 6.42mm existed between the professionals and clerical support staff.There is significant difference in mean PMMT between the Igbos and Hausas. Psoas major muscle thickness varies significantly with occupation, ethnicity, age, gender, and anthropometric variables.

Index Terms— Sonography, Evaluation, Apparently, Healthy, Psoas muscle, Thickness

2 INTRODUCTION

The Psoas major muscle is a long, thick, fusiform muscle which lies lateral to the lumbar vertebrae (Moore et al, 2014). It is the only muscle that connects the lumbar spine and the lower limb (Arbanas et al, 2009).

The knowledge of human muscle architecture has until recently been based on the dissection of cadaver specimens or biopsies and this has been grossly unreliable due to the shrinkage of embalmed cadaver muscle tissues (Quint et al, 1998).The various imaging techniques used for the assessment and evaluation of the muscle thickness include magnetic resonance imaging (MRI) (Hides et al, 2008; Valentin et al, 2015) and computerized tomography scanning (Hasselager and Gogenur, 2014); however, these methods are very expensive.

Recently, there is a growing interest in the use of real-time ultrasound imaging to evaluate the abdominal muscle structure and function in studies on healthy individuals (Hildenbrand and Noble, 2004).The high acoustic contrast of muscle with the adjoining tissue makes it particularly suitable for ultrasonographic evaluation (Moseley et al, 2002). Muscle thickness (MT) measured by ultrasound has been widely employed to evaluate the changes in muscle size occurring because of resistance training-induced muscle hypertrophy (Abe et al, 2000; Starkey et al, 1996) and age-related muscle atrophy (Abe et al, 2014; Sanada et al, 2007). Takai et al (2011) reported that ultrasonographic measured psoas major muscle thickness, strongly correlated to MRI-measured psoas major muscle cross-sectional area.

The reliability and reproducibility of ultrasound as a good technique for measuring psoas major muscle thickness by practitioners of varying experience, especially in areas such as radiographic anatomy, musculoskeletal therapy and sports physiotherapy has been previously established (Petrous, 2002).Muscle thickness measured by ultrasonography has been shown to be a predictor for estimating the cross-sectional area (CSA) and/ or volume of limb muscles (Ishida et al, 1995; Miyatani et al, 2002;Miyatani et al, 2004; Akagi et al, 2010).Psoas major cross-sectional area has been used in several studies to predict the total body lean muscle mass (Shen et al, 2004), sarcopenia (Jones et al, 2015), and surgical outcomes in elderly patients (Durand et al, 2014; Englesbe et al, 2010). It therefore represents a potential marker for cardiorespiratory fitness as it can be used to predict maximal oxygen uptake (VO2max) albeit in men only (Thomas et al, 2017).

To the best of our knowledge, no reference values of the psoas major muscle thickness has been established and no ultrasound-based evaluation of the thickness of the psoas major muscle among the study population has been reported in the literature, hence the need to carry out this study.

3 MATERIALS AND METHODS

This was a prospectivecross-sectional study conducted in a private diagnostic Centre in Port Harcourt, Rivers, state. Atotal of two hundred and ten (210) apparently healthy volunteers was recruited for this study using a purposive sampling technique.Participants were within the age range of 18 to 80 years and were those who came for ultrasound investigation during the study and met the inclusion criteria.Ethical approval for the study was obtained from the ethical Committee of Faculty of Health Sciences and Technology, NnamdiAzikiwe University, Nnewi Campus, Anambra State. Informed consent wasthen obtained from the participants and the procedure was explained to them.

Individuals who had history of trauma, surgery or inflammatory diseases involving the Psoas major muscle and pregnant women, were excluded from the study.Volunteers with marked increased abdominal fat were also excluded due to poor visualization and assessment of Psoas major muscle.

The participants' age, gender, weight, height, and body mass index (BMI) were determined and recorded before the procedure. Height and weight were obtained with a Health Scale, Model ZT-150A. Their height was measured without shoes, heavy outer garments and hair ornaments. The back of the head, shoulder blades, buttocks and heels were touching the stadiometer. They were asked to remove heavy outer garments, handbags, take off their shoes and empty their pocketsto measure their weight.

A 4D LOGIQ P6 General Electrics Ultrasound machine equipped with a 3.5 MHz curvilinear, 7.5 MHz linear, and 7.5 MHz sector transducers was used to examine the subjects. The high frequency (7.5 MHz) linear transducer was used to measure psoas major muscle thickness. The participantslaid supine on the couch since this position facilitates contact between the probe (transducer) and the muscle to avoid having different values of the psoas major muscle thickness in different directions (anisotropic effect), which can occur if the muscle is shortened in a sitting position (Egwu et al., 2017). Acoustic gel was then applied freely at the epigastric, umbilical, and hypogastric regions of the volunteers. The subjects were scanned in a superoinferior approach along the path of the psoas major muscle on brightness mode (B-mode). The 7.5MHz ultrasound linear probe was placed in the transverse plane while the transverse view of the psoas major muscle was obtained. The measurements were taken at the point where the psoas major muscle has its thickest measurement at the level of L4-L5 of the lumbar vertebrae which corresponds to the level of anterosuperior iliac spine (Egwu et al., 2017).Three measurements were made, and the average taken to ensure reliability.

The statistical analysis was conducted using statistical package for social science (SPSS) version 21.0 (SPSS INC, Chicago, Illinois, USA). The one-sample Kolmogorov-Smirnov test and Shapiro-Wilk test were used to test for Normality of the population from which the data was drawn. Two-way ANOVA Tukey post hoc test at 95% Confidence Interval and P values less than or equal to 0.05 was used to test for the relationship between PMM thickness and occupation and ethnicity. Pearson correlation analysis was used to test for the strength and relationship between PMM thickness and Age, Body Mass Index, Height and Weight.Independent-Samples T-test was used to test for significant difference between PMM thickness of male and female subjects who participated in the study.

3. Results

Two hundred and ten (n=210) apparently healthy subjects of different occupational and ethnic background between the ages of 18 and 75 years were recruited for this study.

Table 1 presents the occupational distribution of the participants in the study. The occupations were grouped using the International Standard Classification of Occupations 2008 (ISCO-08) into Managers, Professionals, Technicians and associate professionals, Clerical support workers, Service and sales workers, Craft and related trade workers, Plant and machine operators and assemblers, Elementary occupations, and Armed forces occupations. Students and Retirees who participated in the study were grouped separately. PMMT was highest among professionals (36.77± 5.01) and least among clerical staff (30.35 ± 4.50).

Table 2 shows the relationship between the psoas muscle thickness and various occupation. It revealed that there were significant differences between various occupational groups ($p \leq 0.05$).

Table 3 revealed that ninety-five (45.24%) of the population were Igbos, twenty-six (12.4%) were Hausas, fifteen (7.1%) were Yorubas and seventy-four (35.24%) from minor ethnic group. A significant difference existed between two ethnic groups ($p \leq 0.05$).

Table 4 presents the sociodemographic characteristics of the subjects studied. One hundred and twenty-six (60%) of the participants were young people between the ages of 18 and 39 years, 66 (31.4%) were middle aged people between the ages of 40 and 59 years and 18 (8.6%) were participants of 60 years and above. Ninety-eight (46.67%) of the participants were males and one hundred and twelve (53.33%) were females. The males had a mean psoas major muscle thickness of 35.9168mm and females had a mean thickness of 32.8348mm.

Table 5 shows the relationship between the mean thickness of psoas major muscle (mm) and other anthropometric parameters of the participants.There is a positive correlation between height and mean PMM thickness and a significant difference existed between them($r=0.234$, $p=0.001$), weight and mean PMM thickness ($r=0.535$, $p=0.000$), BMI and mean PMM thickness ($r=0.443$, $p=0.000$).

DISCUSSSION

Findings from this study shows that the mean psoas major muscle thickness is 34.28±5.15mm (range: 20.15-49.90mm).The mean psoas major muscle thickness obtained from the present study is within the range of mean psoas major muscle thickness obtained from the study of Egwu et al, (2017)in adult male subjects when they evaluated occupational-based activity-related differences. The present study revealed that there is adifference in the psoas major muscle thickness between Professionals (Engineers, Surveyors, Lecturers/Teachers, Architects and Radiographers) and Clerical support staff (relatively sedentary workers), Professionals and Elementary occupation (mainly cleaners and labourers), Technicians and Associate professionals (Technicians, Aviation staff, footballers, Offshore staff) and Clerical Support Staff and Plant(Machine Operator & Assemblers) and Clerical Support staff. These observed mean differences between these occupations could be because of occupation- related physical activities on the muscle that provide stimuli for combination of mechanical and metabolic stress which appears to be a potent stimulus for change in muscle size and strength increase (Toigo and Boutellier, 2006). Occupation-based activity-related differences in Psoas major muscle thickness in adult male subjects was also noted in a study done by Egwu et al (2017).

In this study there is significant difference in mean psoas major muscle thickness of two of the ethnic groups. This difference existed between the Igbos and Hausas and this shows that ethnicity may affect mean psoas major muscle thickness. The findings of this study on the relationship between psoas major muscle thickness and ethnicity are affirmed by the study done by Hanson et. al., (1999) on anatomical differences in the psoas muscles in young black and white men using forty-four (44) fresh male cadavers of similar age,weight, and height.

This study revealed that age affects the size of the Psoas major muscle thickness and is highest among middle aged subjects between the ages of 40-59 years with a mean thickness of 36.69mm and a range of 22.55mm (49.90-26.45mm). Previous study by Imamura et al, (1983) noted that age influences the dimensions of the psoas major significantly. However, findings from this study contrast with the that of Egwu et al, (2017) who observed no correlation between age and thickness of the psoas major muscle (r=0.159, p>0.05). This, they alleged could be a result of the narrow age range used for their study. However, the observed age-related difference in psoas major muscle thickness among the study group was not distinctive neither did it follow an exact trend as seen in the study done by Imamura et al, (1983) where psoas major muscle was observed to reach maximum anatomical cross-sectional area in men at age 30 and declined to about two thirds of its cross-sectional area at age 40. The observed relationship between age and psoas major muscle thickness in this study maybe associated with occupational activity of the middle-aged subjects as they were mainly Professionals, Plants and Machine Operators and Assemblers, Force men, Service and Sales Workers and Craft and related trade workers (Welders and Mechanic).

The mean psoas major muscle thickness ishigher in males than females as observed in this our present study.This agrees with other previous studies which revealed that men have significantly more skeletal muscle mass than women even after adjustment for body weight and height (Baumgartner et al., 1999; Beneke et. al., 1991; Bevier et al., 1989; Janssen et al., 2000). Imamura et al, (1983) noted that sex influence the dimensions of the psoas major significantly. In muscle quantification studies based on magnetic resonance imaging, the muscle mass of the psoas major showed significant sex differences while the rectus abdominis, latissimus dorsi, or thigh muscles did not (Hoshikawa et al., 2011; Marras et al., 2001).

There is a significant difference ($p \leq 0.05$) between psoas major muscle thickness and other anthropometric parameters such as body mass index, height and weight in this study. Similar positive significant relationship between the BMI and thickness in all muscles of external abdominal layer was reported by Springer et al, (2006). This findingis in contrast with the study done by Egwu et al 2017, where no relationship was observed between psoas major muscle thickness and weight, height, and body mass index.

CONCLUSION

This study has shown that psoas major muscle thickness varies significantly with occupation,

References

Abe, T., DeHoyos, D.V., Pollock, M.L., Garzarella, L. (2000).Time course for strength and muscle thickness changes following upper and lower body resistance training in men and women. Eur J Applphysiol 81: 174-180

Abe, T., Loenneke, J.P., Thiebaud, R.S., Fukunaga, T. (2014). Age-related site-specific muscle wasting of upper and lower extremities and trunk in Japanese men and women. Age (Dordrecht, Netherlands). Age, 36; 813-824

Akagi, R., Takai, Y., Kato, E., Wakahara, T.,Ohta, M., Kanehisa, H. (2010). Development of an equation to predict muscle volume of elbow flexors for men and women with a wide range of age.Eur J ApplPhysiol.108:689–94.

Arbanas, J., Klasan, G.S., Nikolic, M., Jerkovic, R., Miljanovic, I., Malnar, D. (2009). Fibre type composition of the human psoas majormuscle with regard to the level of its origin.J Anat. 215:636–41

Baumgartner, R.N., Waters, D.L., Gallagher, D., Morley, J.E., Garry, P.J. (1999). Predictors of skeletal muscle mass in elderly men and women. Mech Ageing Dev. 107:123-136. (PubMed)

Beneke, R., Neuerburg, J., Bohndorf, K. (1991). Muscle cross-section measurement by magnetic resonance imaging. Eur J ApplPhysiolOccupPhysio. 63:424-429. (PubMed)

Bevier, W.C., Wiswell, R.A., Pyka, G., Kozak, K.C., Newhall, K.M., Marcus, R. (1989). Relationship of body composition, muscle strength, and aerobic capacity to bone mineral density in older men and women. J Bone Miner Res. 4:421-432

Durand, F., Buyse, S., Francoz, C., Laouénan, C., Bruno, O., Belghiti, J, et al.(2014). Prognostic value of muscle atrophy in cirrhosis using psoas muscle thickness on computed tomography. J Hepatol; 60:1151-7.

Egwu, O.A., Eteudo, A.N., Akunna, G.G. and Uchewa, O. (2017). In vivo assessment of thickness of the psoas major muscle in adult male subjects: evaluating occupation-based activity-related differences. Arch MedSci Civil Dis. 2. 41-47.

Englesbe, M.J., Patel, S.P., He, K., Lynch, R.J., Schaubel, D.E., Harbaugh, C., et al. (2010). Sarcopenia and mortality after liver transplantation. J Am CollSurg; 211:271-8.

Gray, Henry (2000). Anatomy of the Human body. 20th ed., Philadelphia: Lea &Febiger, 1918; Bartleby.com, 2000.

Hanson, P., Magnusson, S.P., Sorensen, H. andSimonsen, E.B. (1999). Anatomical differences in the Psoas muscles in young black and white men. J. Anat.194, 303-307

Hides, J.A., Belavy, D.L., Stanton, W., Wilson, S.J., Rittweger, J., Felsenberg, D., et al (2007). Magnetic resonance imaging assessment of trunk muscles during prolonged bed rest. Spine. 32(15):1687-1692. (PubMed)

Hides, J.A., Stanton, W.R., Freke, M., Wilson, S., McMahoo, S. andRicharson, C.A. (2008). MRI study of the size, symmetry and function of the trunk muscles among elite cricketers with and without low back pain. Br J Sports Med; 42(10):809-813

Hildenbrand, K. and Noble, L. (2004). Abdominal muscle activity while performing trunk-flexion exercises using the Ab Roller, ABslide, FitBall, and conventionally performed trunk Curls. J Athl Train; 39(1):37

Hodges, P.W. and Richardson, C.A. (1997). Contraction of the abdominal muscles associated with movement of the lower limb. PhysTher; 77(2):132-142

Hoshikawa, Y., Muramatsu, M., Iida, T., Ii, N., Nakajima, Y. andKanehisa, H. (2011). Sex differences in cross-sectional areas of psoas major and thigh muscles in high school track and field athletes and non-athletes. J physiolAnthropol. 30:47-53

Imamura, K., Ashida, H., Ishikawa, T. andFusk, M.J. (1983). Human major psoas muscle and sacrospinalis muscle in relation to age: a study by computed tomography. Journal of Gerontology 38.678-681.

Ishida, Y., Kanehisa, H., Carroll, J.F., Pollock, M.L., Graves, J.E. and Leggett, S.H. (1995).

Body fat and muscle thickness distributions in untrained young females.MedSci Sports Exerc.;27(2):270–4

Janssen, I., Heymsfield, S.B., Wang, Z. and Ross, R. (2000). Skeletal muscle mass and distribution in 468 men and women aged 18-88 years. Journal of Applied Physiology 89: 81-88

Jones, K.I., Doleman, B., Scott, S., Lund, J.N. and Williams, J.P. (2015). Simple psoas cross-sectional area measurement is a quick and easy method to assess sarcopenia and predicts major surgical complications. Colorectal Dis;17:O20-6

Marras, W.S., Jorgensen, M.J., Granata. K.P. andWiand, B. (2001). Female and male trunk geometry: size and prediction of the spine loading trunk muscles derived from MRI. ClinBiomech (Bristol, Avon); 16:38-46.

Miyatani, M., Kanehisa, H., Kuno, S., Nishijima, T. andFukunaga, T. (2002). Validity of ultrasonography muscle thickness measurements for estimating muscle volume of knee extensors in humans. Eur J ApplPhysiol.;86:203–208.

Miyatani, M., Kanehisa, H., Ito, M., Kawakami, Y. andFukunaga, T. (2004). The accuracy of volume estimates using ultrasound muscle thickness measurements in different muscle groups. Eur J Appl Physiol.; 91:264–72.

Moore, K.L., Dalley, A.F. andAgur, A.M. (2014). Clinically Oriented Anatomy.7th Edition. Philadelphia: Lippincott Williams and Wilkins.
Moore, K.L., Dalley, A.F. andAgur, A.M. (2009). Clinically Oriented Anatomy.Lippincott Williams and Wilkins.

Moseley, G.L., Hodges, P.W. andGandevia, S.C. (2002). Deep and superficial fibers of the lumbar multifidus muscle are differentially active during voluntary arm movement. Spine; 27: e29-36

Petrous, P. (2002).Ultrasound of muscles.Eur J Radiol; 12: 35-43.

Quint, U., Wilke, H.J., Shirazi-Adl, A., Parnianpour, M., Loer, F. andClaes, L.E. (1998). Importance of the intersegmental trunk muscles for the stability of the lumbar spine. A biomechanical study in vitro. Spine; 23:1937-45

Sanada, K., Kuchiki, T., Miyachi, M., McGrath, K., Higuchi, M. andEbashi, H. (2007). Effect of age on ventilator threshold and peak oxygen uptake normalised for regional skeletal muscle mass in Japanese men and women aged 20-80 years Eur J Applphysiol 99:475-483

Shen, W., Punyanitya, M., Wang, Z., Gallagher, D., St-Onge, M.P., Albu, J., et al. (2004). Total body skeletal muscle and adipose tissue volumes: Estimation from a single abdominal cross-sectional image. J Applphysiol97:2333-8

Springer, B.A., Mielcarek, B.J., Nesfield, T.K. andTeyhen, D.S. (2006). Relationships among lateral abdominal muscles, gender, body mass index, and hand dominance. *J* Orthop Sports PhysTher, 36(5):289-97 Springer

Starkey, D.B., Pollock, M.L., Ishida, Y., Welsch, M.A., Brechue, W.F., Graves, J.E., et. al. (1996). Effect of resistance training volume on strength and muscle thickness.Medsci sports Exerc, 28:1311-1320

Takai, Y., Katsumata, Y., Kawakami, Y., Kanehisa, H. and Fukunaga, T. (2011). Ultrasound method of Estimating the cross-sectional area of the Psoas major muscle. Journal of the American college of sports medicine

Thomas, E.L., Fitzpatrick, J., Chambers, E.S., Parkinson, J.R.C., Frost, G. and Bell, J.D. (2017). Psoas major cross-sectional area.A potential marker of cardiorespiratory fitness. International journal of clinical andexperimental physiology

Toigo, M. andBoutellier, U. (2006). New fundamental resistance exercise determinant of molecular and cellular muscle adaptations Eur.J.Appl. physiol.97:643-663 [PubMed]

Valentin, S., Licka, T. and Elliott, J. (2015). Age and side-related morphometric MRI evaluation of trunk muscles in people without back pain. Man Ther.; 20(1):90-95

Table 1: PMMT according to occupation

Occupation	Frequency (%)	Mean ± STD
Managers	15 (7.1)	35.08 ± 5.063
Professionals	41 (19.5)	36.77 ± 5.01
Technicians & assistants	18 (8.6)	35.93 ± 3.25
Clerical Support staff	22 (10.5)	30.35 ± 4.50
Service and sales staff	31 (14.8)	33.69 ± 4.67
Craft and related trade staff	16 (7.6)	33.60 ± 6.08
Plant and machine operators/assemblers	11 (5.2)	36.55 ± 3.01
Elementary occupations	18 (8.6)	31.53 ± 4.89
Armed forces	11 (5.2)	35.18 ± 4.73
Retirees	8 (3.8)	35.21 ± 2.83
Students	19 (9.0)	32.73 ± 5.09

Table 2. Relationship of psoas muscle thickness with occupation

Occupation	Mean Difference	p value
Professionals Tech. & associate professionals	-6.4200	.000
Tech. & associate professionals Clerical support staff	-5.5750	.014
Clerical Support Staff Plant & machine Operators & Assemblers	-6.2046	.023
Elementary occupation Professionals	-5.2367	.007
Clerical support staff Professionals	6.4200	.000
Professionals Elementary Occupation	5.2367.007	
Tech. & associate professionals Clerical support staff	5.57500	.014
Plant & machine Operators & Assemblers Clerical support staff	6.2045	.023

Table 3: PMMT according to Ethnic groups.

Characteristics		Frequency (%)	Mean Thickness	Mean Difference	P value
Igbo		95 (45.2)	34.7 ± 4.51		
	Yoruba			-.42357	.991
	Hausa			3.1670	.027
	Others			.36383	.967
Yoruba		15 (7.1)	35.18 ± 6.37		
	Igbo			.42351	.991
	Hausa			3.59051	.131
	Others			.78734	.947
Hausa		26 (12.4)	31.51 ± 6.47		
	Igbo			-3.16700	.027
	Yoruba			-3.59051	.131
	Others			-2.80317	.076
Others		74 (35.2)	34.39 ± 4.91		
	Igbo			-.36383	.967
	Yoruba			-.78734	.947
	Hausa			2.80317	.076

Table 4: Socio-demographic characteristics of the subjects and

Independent t test for males and females

Characteristics	Frequency (%)	Mean PMMT	Range	P value
Age (years)				
18-39 (Young)	126(60)	32.82 ± 4.959	20.15-43.65	
40-59 (Middle)	66(31.4)	36.69 ± 4.568	26.45-49.90	
60 & above (Old)	18(8.6)	34.93 ± 3.965	27.70-44.70	
Gender				
Male	98(46.7)	35.92 ± 5.045	21.98-49.90	
Female	112(53.3)	32.83 ± 4.796	20.15-42.90	0.0010

Table 5. Relationship of PMMT with Anthropometric parameters

Parameters	correlation coefficient(r)	P value
Age (Years)	0.304	0,000
Height (m)	0.234	0.001
Weight (Kg)	0.535	0.000
BMI (Kg/m^2	0.451	0.000

Fig.1. Sonogram showing how measurement was taken

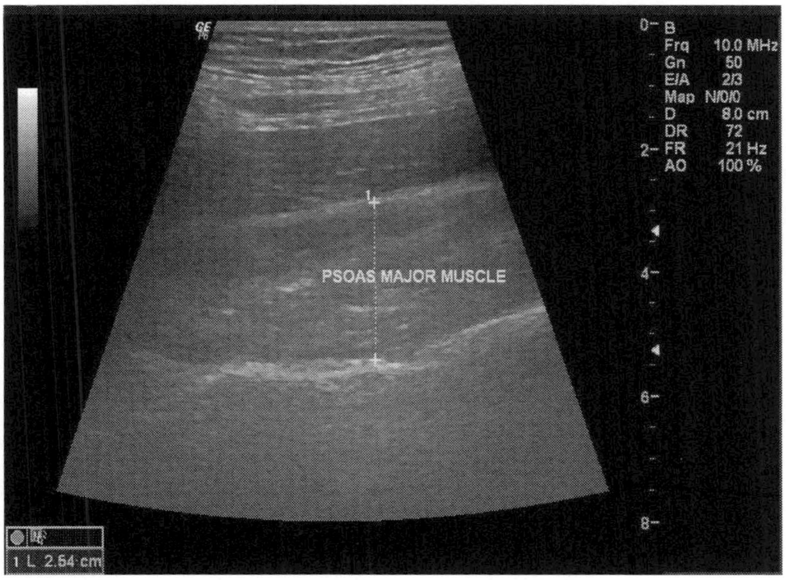

Made in United States
Orlando, FL
04 November 2026

90201383R00020